The 5 Keys to a Clear Mix

Create YOUR Mix Philosophy
for
Christian Artists, Songwriters,
and Church Sound Mixers

STEPHEN ROBERT CASS

songs4God.net Media
PHOENIX, ARIZONA

songs4God.net Media

SONGS4GOD.NET MEDIA
An imprint of Solid Walnut Music
15620 S. 14th Place
Phoenix, AZ 85048
songs4god.net
Send feedback to feedback@songs4god.net

Printed in the United States of America
10 9 8 7 6 5 4 3 2 1
Library of Congress Cataloging-in-Publication Data
Cass, Stephen Robert
the 5 keys to a clear mix: create your mix philosophy for
christian artists, songwriters, and church sound mixers
by Stephen Robert Cass.
Includes bibliographical references.
1. Cass, Stephen Robert—Audio. 2. Recording Studio—Mixing—Mastering
3. Audio—Sound Reinforcement

ISBN 978-1-7378891-4-4 (pdf)
ISBN 978-1-7378891-5-1 (epub)
ISBN 978-1-7378891-3-7 (pbk)
ISBN 979-8-9855371-1-6 (audio)

Copy and line editing, proofreading:
Pauline Goff and Mandy Williamson of Fresheyesproofreaders.com
Cover design: SelfPubBookCovers.com / RLSlather
Interior layout: Formattedbooks.com
Printing in the Unites States: IngramSpark and Amazon KDP

CONTENTS

Titles by Stephen Robert Cass

Fishing in Church: How to Be a Congregational Songwriter
A blueprint for learning the craft of congregational songwriting and getting your songs heard

The 5 Steps to Get Your Songs Heard:
A Congregational Songwriting Plan
Bust All Fantasies that the Music Industry will Save You

Establishing a Culture of Lead Worshipers:
How to Build a Worship Team
Everyone on the platform is a lead worshiper

The Harmony for Worship Project
Training Voices to Praise the Living God

Worship Songs and the Law
How Churches Stay Legal and Songwriters Get Paid

The Proverbs 27.17 Song Critique Method
The Power of Group Learning to Deliver Songs

For John Madrid, the best worship leader in Heaven
who needed someone like me on Earth.

INTRODUCTION

Another book on mixing?

I know. Here's another book on mixing audio from someone who has experience. This book is unique, though. How so? It's a clearly communicated primer on the art of mixing for the Christian artist, songwriter, and church live sound mixer. I take a complex subject and simplify it.

But wait, you say. These people do different things with sound. They might. But what each has in common is the need to mix multiple tracks in a stereo landscape of sound in order to produce a two-track or live presentation for their fans, a demo, or a congregation. The final product for each is a pleasant and accurate sound reinforcement creation. Worship leaders and pastors: this is an excellent starter for the people running your soundboard. Songwriters and artists: this is a smart primer for creating song demos and understanding the language of recording and multi-track mixing consoles.

I'm not going in depth for any category; I'm bringing a simple mix philosophy that applies to all. There are five fundamental principles within sound mixing that are critical to an audio presentation. You need to get on the path to discover your *own* philosophy of how to integrate these 5 Keys to mix your demos, records, and live settings. You become a confident soundscape artist when you get to this point.

I don't have any Grammy Awards. Sorry. But I do have over 50 years of singing, playing, leading, and mixing audio in the church, studio, and other live settings, and I have recorded and produced 14 studio albums, with hundreds of additional hours in the studio creating tracks and demos. I've been crafting live and studio sound since an incredibly young age. That's something.

My first record credits came on vinyl in 1986 as a musician and assistant producer with the Pat McManus Band, https://www.discogs.com/release/5444391-The-Pat-McManus-Band-Your-Home-Town. Here is the rest of my discography: https://songs4god.net/discography.

I promise not to pontificate and offer sacred opinions, nor try to create new jargon for the knowledge of mixing, but to offer a concise launch point for your learning journey. The *5 Keys* method is a foundation so that you can build your own mix philosophy and start feeling confident with your mix decisions.

The road ahead

There's a ton of information written on this subject, but nowhere have I found it wrapped up in a simple philosophy. And you want a simple philosophy starting out—or to enhance your skill set to see if you're on track.

Save your effort of hanging out in forums and scanning hundreds of videos and combing through books—when so many wish to show you how smart they are or talk about concepts that are foreign to you—and let these pages show you how to magnify this deep creation desire to build beautiful music for others to enjoy.

This is an easy-to-read book that you can finish in one sitting. It should take you a little over 45 minutes. But I challenge you to take your time with it and refer to it often.

Most importantly, I want to light the torch and then pass it to the professionals that I trust. These folks are excellent communicators with king-of-the-hill philosophies that can't be beat. I'll give you specific recommendations on these professionals in the *5 Key References* section.

I wanted to write this book because of my passion for you as a Christian recording artist and sound mixer, and I have a deep craving to help the congregational songwriting community. I love sharing, and I'll show my clear philosophy of mixing in a way that communicates that *it's all about the listener.*

God is good. And my way of giving glory to him is by giving a simple picture of the mixing process so you can eliminate FOMO (the fear of missing out), get an easy grasp on the subject, and spend your time increasing your ministry instead of being weighed down by the technical aspects of sound and wires.

I've written this book for Christian artists, songwriters, and church mixers, but I designed this philosophy for anyone seeking a strong start in the recording and mixing world. You'll find simple truths about the mix process that will skyrocket your understanding and set you on the path to success with your listeners, fans, and congregations.

CHAPTER ONE

10,000 ft. View

This book will show the five fundamental pillars for sound reproduction, so you present a clear mix of your music to make great records and mix live audio.

In the art of producing commercially viable music, and by that I also mean for a congregation that thrives on heartfelt worship songs, there are five key areas in the mixdown process where you have *direct control* over the listener experience. Your attention to detail in these areas guarantees the attention of your listeners.

How? Well, working with a great song is optimum, for starters. With these processes in mind, you'll present it in the best possible light and remove sonic distractions. That's what this book is about. A grasp of these five fundamentals in mixing will help you understand what's needed to present a multi-track project and expertly reduce it to two tracks.

But that's only the beginning. This isn't a full guide on the mixdown process. *This is an essential encapsulation of the basics, which many books on mixing pass over. When the basics are mismanaged, mix decisions can kill songs.* This is a method you can use to chart a successful course and create your best work.

After reading this book, read and view as much as you can about mixing and the mixdown process from others. But read everything with a grain of salt and learn to use your judgment. There's no magic out there in mix land. Ask yourself if the information makes sense considering the mixing fundamentals you already know. Don't adopt a practice you don't understand. That's what I hope you walk away with after you read this.

There are many mixdown philosophies, and I'll refer you to five of the best mixdown masters on the planet who have created their own. In the end, take these methods and distill from them the best of the best. Make them *your own*. Everyone will have tips and tricks that they use. If you like them and they make sense for your situation, incorporate them into your own mixdown philosophy.

The difficulty can be that many of the philosophies you'll read about on the internet are advanced, quacky, or merely opinion. Beware, for it seems this genre is full of people who are misinformed, full of anger, and lazy. There is also wonderful input from those with great experience, but they're often drowned out by the former. Nothing you read or hear is mix "gospel" until it aligns with your personal sensibilities. You'll often find a collection of tips from well-meaning mixers who expect that you already have your own basic mixing philosophy.

So, that's what I'll help you build. I'll show you how to create your own philosophy and confidence in making mix decisions. You'll develop a picture of the basic mixdown procedure in your mind so that you can give your project your best effort *now*. The musical stage and the players will appear before your mind's eye. The faders and knobs before you become your paint brush. Their musical offerings, your colors.

The 3-D picture is painted as you walk through a house with five rooms. Unlike a proper house, you can walk through the walls to get to the other rooms, yet each room is distinct. Each room supports the house, and your goal is to let all of the voices and instruments live compatibly in the five rooms, yet each room remains a unique part of the whole.

Your mastery of this 3-D image will help develop your own artistry as a mixer. You'll begin your journey with confidence, knowing you are developing a strong philosophy that will serve you and those around you now, and for all future projects.

The 5 Keys

You've worked hard and given your heart and soul to your recorded tracks, or you're counted on to make the best mix from the live channels on a soundboard. Now, take your project to the next level during the mixdown process and make it stand out to your target audience.

Your goal is to build a great sounding mix built on the strengths of the tracks present. Once you get through one song, you will have a template that you can use with confidence on any song.

By mastering the *5 Keys to a Clear Mix*, you will be well on your way to achieving success as a great sound mixer. Picture the *5 Keys* methodology as the fundamental rooms in a 3-D mixdown house, where each room is labeled like this:

- Mix Space
- Stereo Pan Space
- Volume Space
- EQ Space
- Effects Space

Imagine seeing the artist or the band in front of you. Even if the actual band *is* in front of you, see them and you in this house. Imagine the sound they produce coming toward your ears, from left to right and right to left, from the back of the stage, bouncing off all the walls, past your ears, and to the audience. See the sound from each voice or instrument belonging in each room of this house. Each voice is present in all five rooms at the same moment.

Now move out of the imagination. What you have in front of your eyes is a stereo sound field. Among this field of sound in front of you, ask yourself *how* each track *should* be represented by each of the five rooms. What is the relationship of the **drums** for the *Mix Space* room of the sound field? Where do they belong in the *Stereo Pan Space* room of the sound field? What is their role, priority, or hierarchy in the mix within the *Volume Space* room of the sound field? Can they be heard, or are they drowning out anything in the *EQ Space* room of the sound field? When is any sound effect on their tracks a value-add or a detractor for the *Effects Space* room of the sound field during the mix?

Repeat for each instrument or voice in the mix. Understand that your decisions depend on the relationships between each instrument or voice and how you want the overall mix to sound.

Fig. 1
My band *Spilled*

These five key areas are critical to the base of any audio mix. The five rooms are the rock-bottom requirements to take a multi-track mix and present a 3-D sound and a mind's eye movie of the song to the listener.

There are additional considerations and ways of sweetening the mix for commercial releases. You will find excellent ideas on this and more from gifted professionals in the *5 Key References* section at the end of this book.

Two bonus topics

There are two bonus topics, *Find the Groove and Build the House* and *Dynamic Mixing*. These bonuses are essential to your final mix. They are indispensable in building a firm foundation for your five-room house. These areas support, amplify, and secure your mix decisions. They will make your song shine. More about them after the *5 Keys*.

CHAPTER TWO

Discovery

This is the most important decision you will make for the mix. It's time to discover all the tracks available to you, what tracks you'll eliminate, and the pecking order of the remaining ones. Eliminating tracks isn't always going to be possible in the live and church mix situations, but the same philosophy applies, and it forms the base of your mix.

A mix is not democratic in that all tracks are not treated equally; that is, some tracks and voices will become more important than others. You will find that each track will need different treatment.

Job number one is to set the song—and you—up for success. You need to know which tracks make the song come alive. The wrong tracks can squash a powerful song.

And sometimes it's not just a question of identifying the "right" tracks. That will be a subjective thing. You need to minimize, or get rid of, *confirmation bias* in the mixdown process.

> **con·fir·ma·tion bi·as** (*noun*)
> the tendency to interpret new evidence as confirmation of one's existing beliefs or theories.

Confirmation bias, in the mix world, is when a person too close to the work can't be objective. They can't do what's best for the song. If you are the artist, it's best to get an outside opinion and another set of ears on the song.

If you get input from other band members, you are likely to hear from the bass player that the bass guitar needs to be more up front, the guitar player wants to hear more guitar, the keyboard player says their part needs to be louder, and on it goes. Take each opinion for what it is, an opinion.

The same is going to be true at church when each musician wants "more me" in their stage monitor. That is also a different subject, one that takes you on a tour of understanding "volume wars" and how that can destroy a mix for the rest of the congregation. This is a critical conversation that, hopefully, leads the musicians to use in-ear monitors. This is an important conversation to have, but ... some other time. For this purpose, the "more me" situation is "confirmation bias" because everyone wants to hear themselves louder.

But it's not always practical to get a fresh set of ears on the song (yet it *is* in the church mix situation. *You*, the audio technician, are that fresh set of ears, that objective voice. Vote by delivering your best mix.). Still, there's the question of responsibility for the final product. Is this your song, or are you mixing this song for a friend? A client? The congregation?

Consider the responsibility

No matter whose tracks these are, consider the responsibility involved:

> *When the song is in its finished form, it will represent the best it can be. And that's on you, the mixer.*

No matter who you're mixing for, consider the following during the *Discovery* method:

- Pretend you've never heard the song. Throw up all the faders.
- Discover what tracks make the song work.
- Leave tracks that hinder the song as backup tracks.

What makes a song "work"?

What makes a song "work"? Answers abound and opinions are many and varied. Prosody is a large part of good songwriting and what makes a great song. That is, the feel of the music fits the emotion of the lyric. Your job as a mixer is to clarify and amplify this emotion and musical feel. Learn more about prosody from a songwriter's perspective in my book *Fishing in Church*, and other resources listed in the *Tools* section at the end of this book. But for the purposes of *Discovery*, the answers are:

- Which tracks best represent the melody of the song?
- What style of music do the tracks reveal the song to be?
- Which tracks support the main vocal best?
- Which tracks distract from the main vocal?
- Which tracks distract from support tracks (those that support the melody or main vocal)?
- Which tracks take away from the song's prosody?
- What instrument(s) best support the groove?
- What instrument(s) distract you from the groove?

In almost every case, less is more. There are often too many tracks as opposed to not enough. Your job is to determine the pecking order of vital tracks, support tracks, highlight tracks, and tracks that don't belong.

The *Discovery* process is not some final voodoo that makes the song work. It's learning to identify what makes the song tick in the first place. This is of the utmost importance.

Do what you know is right

This exercise will be tough! If this song is yours, you've already recorded these tracks that you think are *da bomb*. Or maybe you've produced these tracks with these voices and instruments and imagined how beautiful they'd be together.

It's a bit like slashing a part of you if you find tracks that don't work well. I get that it can be a knife in the heart. I'm a creative too.

All songs that work have one thing in common: There are maybe five instruments/voices that really define the groove. These tracks are usually all that are needed to support the main feature, the vocal. The larger collection of tracks may prove to distract from the clarity of the mix.

When you pretend you've never heard the tracks of the project before now, it's like listening to the song with a fresh pair of ears.

But it might be necessary to slash some tracks.

> ***Note**: I know this can't happen when you're mixing for the worship band. But get a sense of which instruments work well for the song and which ones support the others. Learn how to use tracks and instruments as highlights that appear and fade in and out. See more about dynamic mixing in Chapter 9: Bonus 1 and 2.*

> *One stark reality of church bands is that you can often hear an instrument loudly enough without amplification through the soundboard. With that, all you can do is use the soundboard to amplify their weakest aspect, if that's desirable. What I mean is that maybe you need to make an EQ adjustment and hear the high end of the guitar more, but you don't need to amplify any more of the midrange frequencies from it. Maybe you hear those frequencies more naturally in the room. Get more ideas about this from Chapter 6: EQ Space.*

If you're mixing this song and it's yours, get that second opinion on what you've discovered. Opinions are just opinions, but it'll give you some feedback on what instruments you think the song needs.

If it's a friend's song, it might make them angry if you cut something out. In the end, you need to find balance. Go for what you know makes the song work best, with fewer tracks, or bring in your friend to consult before you do.

Clear communication

If you're mixing sound at church or any live event in a room, your job is primarily sound reinforcement. That is, do your best to faithfully present the sound you hear in the room from the instruments and vocals naturally and reinforce them. It's to clarify one instrument or voice from the rest when it's weaker, or to find and mute a louder sound for the sake of balance. But be assertive. Your goal should be to make the band or artist shine. Decide to be proactive instead of reactive. Be intentional.

You are the person deciding how best to present this song; you set the stage for how the world will hear the song in its final version.

Realize that *you* are in control of the outcome. If the drums are too loud for the room and the mix, you need to tell whoever is in charge so that can be fixed. Without fixing problems like this, it's akin to the old saying, "You can't put lipstick on a pig."

So, approach the mixdown with confidence. No one would have approached you or hired you if you weren't the best mixer in the room or they didn't value your opinion. I realize that many soundboard positions in church are volunteers, and that might be you. But you can impress the worship leader with your knowledge and show them you're a part of the team, not just a warm body.

If it's a client, well, do what you're getting paid to do. Put your best-mix ears on. Discover what tracks work for the song. Create the best final product you can. This is your job. Present your own mix or balance to them *first*. Then let them comment.

If you need to compromise, then compromise. But stick to your conclusions when you discover tracks that don't make the song work. It's your call on which individual tracks make the cut for the song and which tracks fall to the cutting room floor.

No matter if you're a hired song mixer or a church volunteer, a clear communication process is a must. In the church and live setting, get feedback from each player about their monitor mix (if you control that). If your worship leader is picky, you are going to have to please them. They may not be picky, and you need to find opportunities to show them *your* ideas.

You may well find yourself in the situation where your mix decisions are scrutinized and must be approved. Find out as soon as possible to whom you answer. Get ready to explain your mix architecture to that person. Demonstrate your decisions to them. If you don't, your great mix decisions could be ignored and may never see the light of day. Express the value that you add to the songs.

Be thrilled to receive production notes and comments on the mix and find common ground. In all cases, stay in charge of the mix. Present updated mixes to stakeholders with the attitude of moving the project forward, not to satisfy whims.

Now on to the *5 Keys*, the essential rooms of your mix house …

CHAPTER THREE

Mix Space

Get the least crowding of tracks or voices in the stereo sound field for the same event in time. Exact timing of each part, instruments and voices on the same beat, is often essential for a mix. It helps to keep the muddiness out. It keeps the overall mix sharp.

We know and love tight music. Whether we're a mixer or a listener, we hear a song and feel its power when it is mixed well. There's always more than one part on a beat.

Here's a weird curiosity: A band playing live and a studio recording of the same song are two totally different animals. We tend to be more forgiving when we watch a band live. Recording engineers/mixers in the studio tend to overproduce the tightness between instruments and make a band artificially tight.

Do you remember when Princess Leia said, "The more you tighten your grip, Tarkin, the more star systems will slip through your fingers"? Well, this is something like that word picture, only the star systems are instruments or voices escaping the beat.

We like that artificial tightness. A great mix in a multi-instrument recording presents distinct and discernible sounds at the same time event, making it punchy and precise. The sounds are combined, and we perceive more space around the beat to process and enjoy the sounds.

When we see a band live, we get engrossed in what we see and hear. The room or auditorium they play has its own acoustic characteristics that add to the experience. The unique room sound, how they have their onstage equipment set up, where you sit in the room or auditorium, and how the songs are mixed all define the experience. If the room isn't treated for acoustics, sound will reflect off the walls and surfaces. We often hear a muddier mix than we expect. And that's ok for most people because they are engrossed in the show and seeing their favorite performers.

Sure, I've been to concerts where I thought the band played the song *exactly* like the record. And that's pretty impressive, but not all bands do that. No matter. We still enjoy seeing them.

Even so, as the listener, we create the sound stage in our minds when we hear a record. When we hear a record, we somehow expect perfection. We want a perfect representation of what we'd *like* to hear if we saw that band or artist live. We're disappointed if the record doesn't deliver that. Alas, technology has allowed us to create some projects that are *too* perfect.

This time event management is critical for both live and studio mixers to present clarity and not offer distraction to the listener. Certainly, the studio mixer in postproduction has more control over this than the live mixer at the sound booth in church. They have choices, though. The studio mixer can move the drum and bass grooves to a time grid (or the producer had the drummer and bassist on a click track and he or she did an excellent job of staying on the click in the recording session). The church mixer can recognize timing inconsistencies and let the worship leader know so he or she can work with the musicians.

But don't get me wrong. I know this sort of thinking can go overboard in both cases. The mix can start feeling unnatural. There's a limit when dissecting the problem technically and making the mix so tight that it's not human. And there are times when less direct words help you to preserve your relationships on the worship team. Because of our forgiving nature, we often ignore timing differences between instruments in the live setting. But the differences become more critical when listening to a recording. Regardless of your taste, we must give the *Mix Space* respect on a recording *and* a live event because we expect each voice and instrument to be clear. Do your best in the situation.

Oftentimes, less is more for the number of instruments present in a recording, as I stated in *Chapter2: Discovery*. Choose the fewest instruments that define the core group of tracks as the basis for your song. Among the core, locate the rhythm and base chord/melody instruments that drive it. These become the backbone, the skeletal structure on which you paint the rest of the song.

Define the backbone

Of the tracks you've selected from *Discovery*, define the backbone instruments, the key elements of the song or band or those that drive the groove:

- Drums,
- Bass guitar, and a
- Rhythm instrument (such as a guitar or keyboard).

Define a highlight instrument, those secondary instruments or parts that support the key elements. These can be the same instruments, but maybe one or more of them aren't as prevalent in the verse, for example:

- Keyboard,
- Guitar, or
- Trumpet.

Define a feature instrument. Those instruments used for effect or used sparingly, such as any of the previous instruments and more.

Any more components make the mix too busy. Sure, of course great songs exist with more than these. This is just a basic rule of thumb. Again, we forgive this sort of thing listening live as opposed to a recording.

Also, many of you will be just recording straight-ahead rock with all instruments blazing. That's ok, the same rule of thumb can apply. See *Chapter 6: EQ Space* and *Chapter 9: Two Bonuses* later in the book for ideas on how to create room for each part.

Now that we've gone through *Discovery* and *Mix Space*, create a spot for those instruments up on a stage.

CHAPTER FOUR

Stereo Pan Space

Or sometimes this is called *Horizontal Space*. The goal is to have each instrument and voice within their own area on a natural-sounding stage (or any stage setup in your mind as is the case when mixing a multi-track recording).

In the previous chapter about *Mix Space*, you're concerned with instrument event timing to provide clarity in that way. Now with *Stereo Pan Space*, your attention turns to spreading out the instruments and voices along the horizon of the stereo sound field. In other words, use stereo panning to place that voice or instrument either left or right of center. In theory, different voices and instruments should occupy their own space.

Instruments and voices can have similar frequencies. Subtle differences in panning can make a vast difference in relieving a crowded mix. Creating individual positions within the stereo panning scheme can add clarity. Whether live or when mixing multi-tracks, this is a great first step toward creating a 3-D listening experience.

Just imagine where you'd hear each distinctive voice or instrument, left to right or right to left, if the band or performer was playing on a stage. This also includes each individual piece of the drum kit.

In practice, the bass drum and bass guitar are located at the center of the LR pan control to provide unity for the lowest frequencies (also to add more punch for rock songs), as well as the lead vocal which provides the usual focal point of the song. The rest of the tracks are where you want them to be.

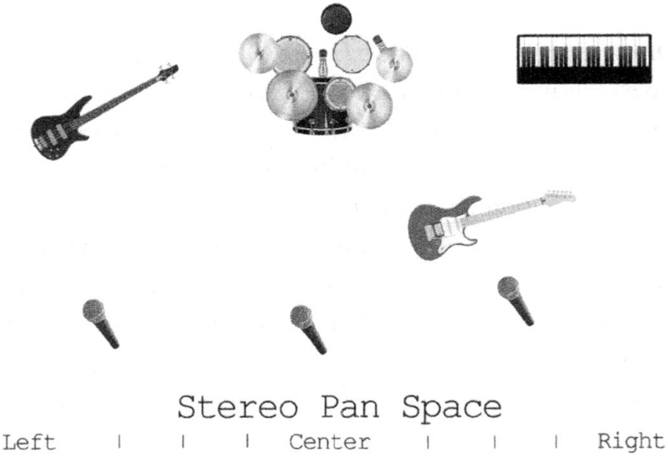

Fig. 2
Stereo Pan or Horizontal Space

Balance the stage

Remember, an instrument/voice may appear predominantly on one side or the other within your mix, so it follows that the remaining percentage of that voice is the other side of the mix. Know how much. Know where this voice is in the stereo panning spectrum.

For example, maybe you want a guitar panned to three o'clock. But when you're facing a stage and listening to a band or a solo performer on the right side of the stage, you'll hear a certain portion of him or her in each ear. Maybe predominantly on the right side, but there'll be a portion going to your other ear.

So, if this is a mono recording of the guitar, you might pan it to three o'clock and then plan on having a reverb returned to ten or eleven o'clock from another channel dedicated to reverb only.

This guitar-on-stage scenario, and where it now resides in the *Stereo Pan Space*, is just an idea for the future of the mix. For right now, you'll just pan that mono guitar channel to three o'clock.

Just be aware that you might want any stereo tracks (guitar recorded by two or three microphones, L and R inputs from a stereo keyboard, etc.) to be only partially panned. Maybe you'll start the mix with L and R panned hard to each side and then re-adjust them as you go. For example, let's say you want the dry tracks of a stereo guitar to appear mostly on the left side. So, you might hard pan one track to the left and leave the track on the right side somewhere around the center.

There are no hard and fast rules other than to be aware of where things reside in the mix.

Get an understanding of how our two ears perceive sounds when listening to a sound scene. Panning will always be a matter of taste in a mix, but it's good to have purposeful control.

Dry vs. wet

At this point in the mix, you're focused on dry tracks (those without effects. I'll also talk about wet tracks, or those with effects, in *Chapter 7: Effects Space*). Remember the difference: dry tracks are those you've discovered from the raw recorded or live instrument/voice tracks.

Wet tracks are those tracks which have sound effects that *you've* added during the mixing process. But they can also be tracks that had effects recorded with the instrument when you originally discovered it. So, those discovered ones with effects might be considered *dry* (just to keep them straight in your mind). In any case, you can't make them *unwet*, if you take my meaning (you can't erase the effect). The point is to manage effects the best you can so they don't muddy the clean mix you've built to this point.

No matter if I reference dry tracks or wet tracks, stereo panning considerations are the same. The sound stage of musicians and instruments in the mind's eye of the listener depends on you providing clarity. Effects can cloud this sound stage in a hurry. Choose wisely where effects are panned in the stereo sound field. More about that when we talk about *Effects Space* in Chapter 7.

CHAPTER FIVE

Volume Space

Volume Space is setting the relative volume of the tracks and is your primary tool for amplifying the depth dimension of your mix.

So far, you've been judicious with the number of voices and instruments by using the *Discovery* and *Mix Space* tools. Because the listener must compete to hear each track, your results help the listener distinctly hear each sound.

Next, you've spread these distinct sounds along the horizon of a sound stage by utilizing the *Stereo Pan Space* tool.

Each adjustment going forward will affect the balances you've achieved thus far. What you adjust will make some other aspect of the mix unclear. You'll have to go back and tweak something to keep chasing after clarity. Count on it.

We hear the world in three Dimensions

In the song, you've tamed the time events (one dimension of the music) and the population of the stereo spectrum (a second dimension) and are beginning to find clarity in the project. With that in mind, there are two aspects of the third dimension you can adjust to add to the palette to paint your sonic masterpiece. The first step of this third dimension is deciding the role of each voice or instrument in the *Volume Space* room. As said earlier, *a mix is not democratic.*

See volume as a straight line from the back of the imaginary or actual stage to the front of your face. This is your depth and relative loudness range. Now make room for each instrument along this line. There is a hierarchy according to which tracks have priority and you'll hear it.

In live mix situations in small rooms, such as churches, the hierarchy may be forced on you. What I mean is that if the drums aren't contained or the room isn't treated well acoustically, they will overwhelm the room and the mix. Short of getting that situation fixed, you will have to learn to live with it. How do you fix that? Get a pillow for the kick drum. Find a short wall or an enclosure for the drums. Ask the drummer to play with hot rods (drums sticks that are a combination of smaller sticks tied together).

Or maybe your situation is that the musician monitors are too loud. The fix is to get the worship band to switch to in-ear monitors. If that's not going to happen, try adjusting the EQ of the monitor signal (see the next chapter) and take some of the low-end out, especially.

It's certainly tough to fix some of these problems. Sometimes church budgets prevent solutions and maintaining relationships is more important. But without addressing issues like these, the mix and the congregation will suffer. Take part in the proactive process of making your worship center ready for live music. Your congregation will be able to enjoy the music and not be distracted by noise.

If the acoustics are right in the live space, you can treat the *Volume Space* adjustments as you would mixing songs in the studio or your laptop. Granted, that's a perfect-world scenario. The old saying is that when life gives you lemons, make lemonade. Do what you can with what you have.

So, in the perfect world, adjust the hierarchy of volume between each instrument or voice in the live or mix room and start giving actual shape to the project. The increase in volume of each track will amplify any previous mix decisions.

The Fletcher-Munson Curve and a Trick

Now that you have the main elements of your mix in their proper places within the hierarchy, it's time to listen to the overall volume of the mix. You will be adjusting and readjusting during the rest of the mix session until you are satisfied that all the elements are in their proper places.

There's a phenomenon with our hearing. As the actual loudness changes, the *perceived* loudness will change at a different rate, depending on the frequency.

This doesn't seem very important in the real world … *until you're the one in charge of mixing a multi-track project!*

It's known as the Fletcher-Munson Curve, and it can be important to your mix decisions. It's great to study. But don't worry about understanding the phenomenon past its basics for now. Understand one trick so you can work with it today.

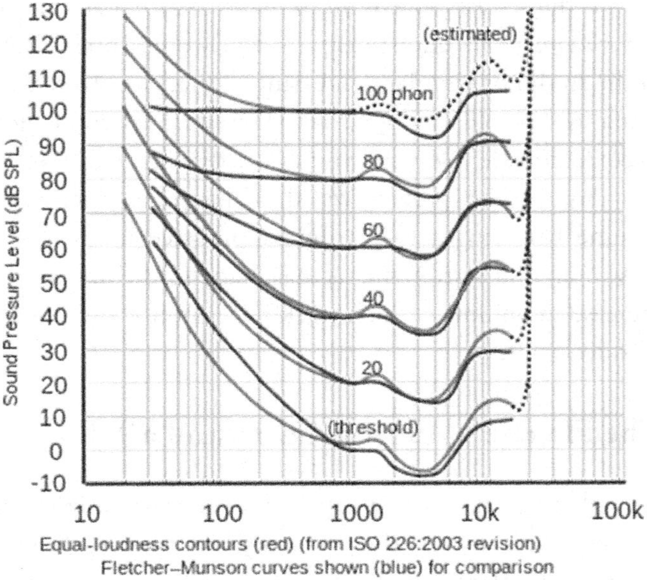

Fig. 3–Fletcher-Munson Curve
Courtesy Wikipedia

The goal with the volume of the overall mix is to hear every part clearly at a very low master volume fader level as well as the intended master volume fader level. The secret:

> *When you turn down the master volume level of the mix to listen for each component and one is missing, just nudge the volume of the missing component upward slightly until you hear it. Then recheck the mix at the desired volume.*

Again, there is no hard and fast rule on this one. Just keep in mind that if a component is unintentionally louder than the primary (e.g., main vocal or lead guitar) then it will be distracting. Repeat this trick as many times as needed until you're satisfied with the overall mix.

My weakness here is that I love to sing. And I love to hear great harmonies. When I mix, I tend to make the background vocals too loud. If it's your desire to have the main vocal above everything else, listen to all vocals in the mix at a very low-volume level to hear that the lead vocal is above the rest.

Another technique along the same line is to start playback and leave the door to your mix room open just a little. Now walk down the hall. Listen for each component while you're down the hall or outside the door to your mixing room. Go back in and adjust each one to the desired volume. This is also a great final mix trick.

Next up is the secret sauce for mixing clarity—*EQ Space*. Studying, knowing, and practicing this art will serve notice to all concerned that you're a serious mixing ninja.

CHAPTER SIX

EQ Space

When you hear the term *EQ*, short for *equalization,* you're hearing that term used as an action to describe the adjustment of the power, or volume, of certain frequencies within a track. Carving the audio on a track in this way is a tool to help it audibly fit with the other instruments and voices in a mix. The unique timbre (that is, the resonance, tone, and color) of a voice or instrument defines its character and helps it stand out in the mix. Recognize these aspects of the voice or instrument as you make adjustments in the *EQ Space* of your project.

Theoretically, each individual instrument or vocal needs to occupy its own space within our range of hearing in the mix. You want them to rub elbows in the mix, but have each distinct voice heard clearly.

Fig. 4. Frequency Chart
Courtesy of audio-issues.com

The chart in Fig. 4 will show you the range of common instruments using piano keys as a guide. Locate the associated frequency range of the notes of the piano in Hz (short for Hertz, the name of the guy who discovered this stuff) at both the top and bottom of the chart.

Each line past the instrument name gives the frequency range of the instrument. The darker first half represents the actual range, while the lighter second half represents the upper harmonic frequencies generated by the instrument. Visit audio-issues.com for a full description.

This chart isn't to overwhelm you with information; the idea is to familiarize you with:

- The frequency range of instruments and voices against the backdrop of the piano
- The challenge audio mixers have in representing and preserving clarity of instruments in a multi-instrument recording
- Strategies to highlight certain instruments and voices

This chart will also be a valuable resource for your library. Here's an example for using the chart when you mix: If the vocal frequency range includes the same lo-mid frequencies as a tom, the result could be mud. The same is true when different vocals compete. Make a separate *EQ Space* for backing vocals as compared with the main vocal by eliminating frequencies of the full range of the backing vocals.

Identifying the EQ footprint of a voice or instrument, and then shaping, trimming, or boosting frequencies within its range, is probably the most important aspect of mixing music.

What is audio?

I need to get complicated in the next few paragraphs because you are entering the audio engineering world. But I'm going to paint a clear picture. Although it is not necessary to understand all the science behind audio to work behind a soundboard, you'll hear the upcoming terms on a regular basis. Now you have a small library of terms and theory for reference. If you'd like, skim this section and begin reading at the *What are EQ adjustments*? section.

What is audio? A source of energy sends its pressure through a medium, such as air. The energy causes the molecules of the air to vibrate. The vibration causes the electrons of the atoms to flow from molecule to molecule. *This transfer of the energy between molecules in the medium is transduction (the conversion of one form of energy to another).* So, the particles of the air transduce the energy and transform it into what our ears perceive as sound.

Take the phenomenon of lightning, for example. The pressure from the bolt through the air causes particles in the air to vibrate, then those particles transduce that energy into something we can hear. The movement of the electrons within the molecules of the particles is where the transduction and transformation processes occur. Moving electrons are the basis of light, sound, mechanical, and electrical energy. The result for sound is our ears perceive that transformed energy into what we call thunder.

An oscilloscope, which is a device used in electronic engineering and repair, gives a visual picture of audio in action. You'll see a similar picture in DAWs (Digital Audio Workstations) and plug-ins.

We can see the electrical sound pressure on an oscilloscope, similar to Fig. 5. It views the voltage, the energy pressure, and how the electrons move through the medium. The voltage pressure causes the electron movement to alternate above and below zero volts. The signal is measured to determine the frequency of this movement, or how often it repeats each second.

Science class is almost over now!

We don't care so much about voltage and the movement of electrons for our topic. But the picture painted in the oscilloscope monitor gives us a visual into the world of adjusting audio frequencies.

The electrical value of sound is quite small. Actual sound pressure sent to a microphone, or the signal generated by a vibrating string into a guitar pickup (also both transducers), is converted into a voltage and current (how we measure electron movement). Tiny amounts of voltage and current! So we can hear them, these signals are then sent to amplifiers and out to a final transducer, a speaker.

Audio frequencies, by the numbers: The higher the pitch, the higher the frequency or cycles per second. Let me paint that picture.

A picture speaks a thousand words

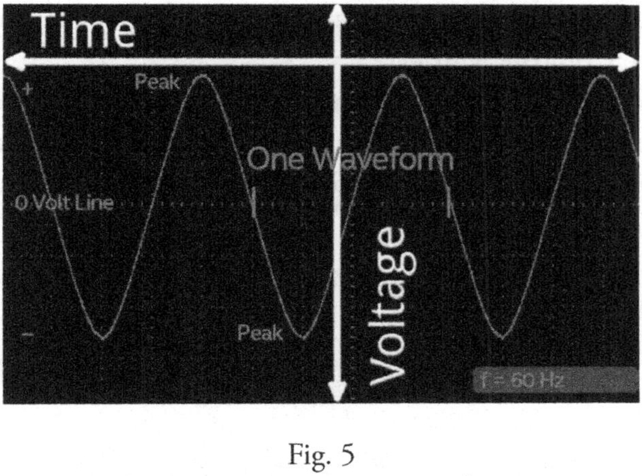

Fig. 5
60 cycles per second on an oscilloscope

We measure the number of cycles in one second and call that the frequency. The frequency of the alternating current is measured as speed in time in which the signal travels (how fast it travels as the signal alternates above and below the zero-volt line).

The faster the speed of the signal—the closer the waveforms are together—the higher the frequency. The shorter the wavelength of the audio signal and the closer together and more frequent the waveforms become, the higher the pitch.

One cycle, or one waveform, is measured as a signal beginning at the zero-volt line. The signal then travels above or below the zero-volt line and then returns to the zero-volt line. This cycle finishes as it repeats in the opposite direction and then returns again to the zero-volt line.

Imagine viewing the outline of a symmetrical mountain. At the right edge of the base of the mountain, imagine viewing the reflected image as if you were looking onto the lake at the base of the mountain. Only the mountain in the lake starts at the right edge of the mountain on the land. Compare with Fig. 5.

If the audio signal sent to the oscilloscope was a constant 60 Hz, then we'd see 60 mountain peaks above and 60 mountain peaks below the zero-volt line per second.

> ***Tip:*** *You've heard of 60-cycle hum? No doubt. Our power from the electric company is delivered to us as alternating current at 60 Hz (in the US). Sixty-cycle hum happens when there isn't proper shielding of audio cables. That 60 Hz signal from the power cable leaks through the air into our audio cables. That signal blends with the good stuff and is amplified and heard on our speakers.*

Noise

Noise is a random combination of unrelated frequencies that have no definite pitch. White noise is a blast of all frequencies with equal power behind it. This contrasts with a musical sound that has a definite pitch, a fundamental frequency, and a combination of related frequencies called harmonics.

Sixty-cycle hum, although we call it noise, is actually a fundamental frequency because it's constant. It will generate harmonic frequencies at 120, 240, and 480 Hz (one, two, and three octaves) as well as other harmonics at symmetrically mathematical intervals.

Often, you'll hear about pink noise, all the frequencies of the audio spectrum, 20 Hz to 20 kHz (20,000 Hz or 20 kilohertz, abbreviated kHz). We use a pink noise generator in testing and calibrating speaker systems, audio equipment and acoustically designed structures (control rooms, studios, auditoriums, etc.)

Pink noise generators are a useful tool because the power behind the noise generated is controlled for each octave, more power at the lower frequencies and less so at the higher ones. Pink noise tests analyze the frequency response of a room to know what frequencies are artificially amplified when they bounce off objects. But it's also great for understanding what frequencies in a room *aren't* heard because they're absorbed.

White and pink audio noise are oh, so random! They're not organized or definite sounds like that coming from a human voice or an instrument.

In practical terms ...

When you hear a voice or an instrument, you're hearing a complex collection of frequencies. You'll see them on the oscilloscope in Fig. 6. It'll contain frequencies within the range of instruments in Fig. 4.

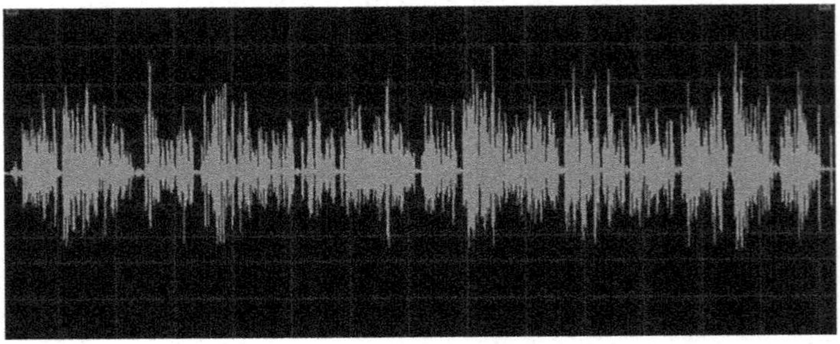

Fig. 6
Complex Audio Waveform

The human hearing spectrum

The audio spectrum is 20 Hz to 20 kHz. This covers the entire range of human hearing. The upper limit of the spectrum will vary, depending on age and gender. The upper limit of the average male is 16 kHz and the average female will be about 18 kHz.

What are EQ adjustments?

Making and understanding EQ adjustments (adjusting the volume gain of frequencies), whether live or in the studio, is what makes you a knowledgeable and sought-after sound engineer. When you hear the term EQ adjustments, you are carving the raw audio recording of the instrument or live sound. You're not creating any new audio but are adjusting the gain of certain frequencies in the complex audio signal. You will find similar methods of adjusting the EQ of each channel on any soundboard or DAW.

The first step into this world is to see the handles of the power adjustments of the frequency ranges in Fig. 7 (the colored circles [black and white here, but full color at their website]). The level is being amplified or reduced as you move them up or down, respectively. Find out what frequency range you're adjusting by seeing the spectrum numbers at the bottom of Fig. 7 (10, 50, 100, 200, etc.).

Pause. Breathe. Re-read. But know you don't need to understand the above to just get in there and start adjusting and listening to the results. Let your ears be the best teacher of them all.

So, you use the EQ tool of Fig. 7 to reduce or amplify the audio at those frequencies. You "cut" or drop (or increase) the gain of those frequencies, measured in decibels (dB), at the colored-circle adjustment points.

The bottom line is that you're making volume adjustments for certain frequencies. This is an excellent tool for removing unwanted noise or frequencies, such as those that cause positive feedback ringing. The bad news is that you'll remove desired audio at the same frequencies.

Your turn

Here are EQ adjustments in action. And I want to give *you* a chance to take this action!

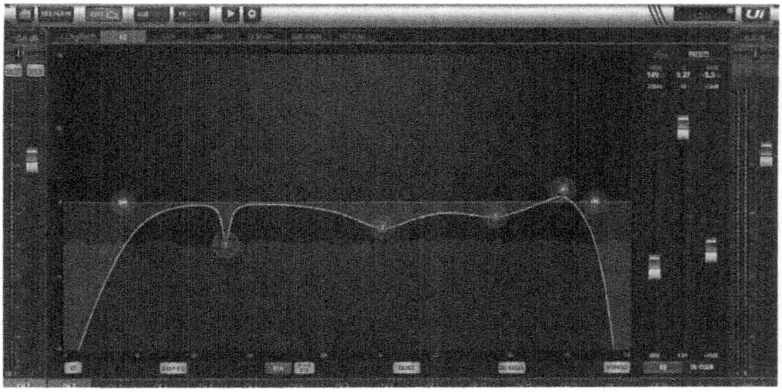

Fig. 7
Parametric EQ
Courtesy of Soundcraft.com

Fig. 7 is a picture of a parametric equalizer on the Soundcraft UI series mixing board. You are watching the EQ gain picture of the signal plugged into Channel 1.

Take a tour of this graphic. You see white/gray wash with the jagged top in the lower half. That represents the incoming signal to the channel. This is a representation of the complex audio signal from Fig. 6. The purpose is so you can visualize the strength, the peak voltage, of the incoming signal against the horizontal line in the middle of the graphic. The horizontal line represents the gain level for the channel.

See the numbers 20 to 22k along the bottom of Fig. 7 and from left-to-right (but above the "CH" numbers). This represents the full sonic spectrum, 20 Hz to 20 kHz. The curved line in this graphic *represents the gain (and control of the gain) of the sound in this channel. EQ adjustments are varying the gain of selected frequencies of the signal within this channel.*

Now, take this action and follow along with the next few paragraphs. Try some of your own adjustments at their product website: https://www.soundcraft.com/ui24-software-demo/mixer.html. When you get there, click on **EDIT** in the upper left, then click on the **EQ** tab below it. Now, select **CH 1** in red on the bottom left.

The first thing I'll say is that you'll love this graphic on their website. If you're anything like me, seeing the colors (and getting my hands on it) will make much more sense of it all.

- To repeat, the white/gray line splitting the image horizontally is the gain level of the channel.
- From left to right, the blue, green, red, purple, yellow, and orange circles represent points of EQ adjustment along the audio spectrum that you can make to the incoming audio of that channel.
- Any adjustment below the white/gray channel gain line is called *subtractive* EQ, and above it, you guessed it, *additive* EQ.
    - The three sliders to the right: The settings of these shown depend on the circle. The circle on the far left is the blue HPF, the high-pass filter and the circle on the far right is the orange LPF, the low-pass filter. The blue and the orange only offer one slider—the frequency setting. This sets the frequency point where gain reduction begins. Adjust by either clicking and dragging on the circle or by adjusting the slider. You'll see three buttons on the right under Slope—12, 24, and 36 dBu—which adjust the sharpness of the frequency drop-off.

- When you click on any other circle, you'll see the three sliders on the right. That frequency set point circle is highlighted and shows the associated sliders to the right:
    - The frequency of the point of adjustment
        - The frequency can be adjusted along the entire spectrum, 20 Hz to 20kHz.
    - The *Q point* of the frequency to be adjusted
        - The Q is the sharpness of the cut around the frequency. For example, the green #1 adjustment is a deep-cut notch filter, while the red #2 is a notch filter with gentler slopes that encompass more neighboring frequencies.
    - The gain of the point of adjustment
        - This red-knobbed slider is the gain above or below the set gain level of the channel.

Adjust the EQ for my vocal

For example, let's pretend to mix my voice in this channel.

Set the High-Pass Filter, represented by the blue HPF circle on the left, so it cuts off any frequencies below 50 Hz presented by my voice in the recording or the live sound amplifier. Do the same for the opposite end of the spectrum, the orange Low-Pass Filter.

- Click and drag the blue circle to the approximate frequency location.
    - Move the circle around as you watch the Frequency slider move on the right side.

- Click on any of the SLOPE buttons on the right side.
- Repeat for the orange circle, the LPF, on the far right.

The green #1 filter is a steep notch filter. I have it set to reduce the gain of the frequencies between 150 Hz to 180 Hz.

- Click and drag the green circle to the approximate frequency location.
- Adjust the Q slider to steepen/lessen the curve.
- Grab the Gain slider to decrease/increase the gain of the frequency.

Repeat for the red, purple, and yellow frequency settings.

I have the red #2 filter set at a gentler slope of a notch filter. This will reduce the gain from the range of frequencies between 500 Hz to 2kHz from my live or recorded voice.

It's a similar slope on the purple #3 notch filter, set to reduce the gain of the 3kHz to 7kHz frequencies.

The yellow #4 notch filter has gentle slopes (lower Q number) and is ever so slightly boosting the frequencies around 11.5 kHz. But the gain at the right edge of this filter drops sharply because …

I have a Low-Pass Filter, the orange circle, set to cut frequencies sharply at 12 kHz and above. And this setting overrides any other that nestles up to its boundary.

Homework

Go to that web page and adjust the settings as I described for adjusting the EQ on my voice in the previous section, which is also shown in Fig. 7, then come back here. Play around with the knobs. Click and drag the colored circles. You won't break anything. Go ahead. I'll wait ...

Oh, I almost forgot. In that graphic, click on the green RTA, *real-time analysis*, button towards the bottom. Now you can see that white/gray wash below the horizontal line representing the incoming signal to the channel. If you were using their actual sound mixer, this is where you'd *see* the incoming signal to the channel as it happens. That is valuable information as you mix.

So, as you experiment with this graphic, you won't be adjusting any audio. But it will give you a feel for what I'm talking about. Click around on any of the other buttons of the online Soundcraft mixer. These are the same basic functions you'll find on any soundboard or DAW. Take some notes for areas you want to learn. It's great practice, and it will prepare you for anything.

Come back to this section of the book often to learn more about this or any DAW. This will give you a great boost to the wider world of EQ. There's much to discover. Visit the websites of the professionals I give in the *5 Key References* section to get their take on EQ moves.

Check out the Soundcraft product page for this mixer: https://www.soundcraft.com/en-US/products/ui24r. It's a good wireless mixer, and you can access and control the screen shown above in Fig. 7 with your iPad, Android, desktop, or laptop. It's a brilliant solution for your band or for church.

Sub-ranges of the Spectrum

To make gain adjustments (equalization) to the sound, it's important to recognize the different parts of the audio spectrum.

Audio Range Description

- In the frequency range of 10 kHz to 20 kHz (top of the range), there's not much energy or musical content. We perceive this as a soft hissing sound. If your mix seems to lack brightness or clarity, don't think about increasing the level of this frequency range, as the only result will be adding lots of unwanted noise (or with extreme power, it could destroy your tweeters or high frequency elements of your speakers).
- The range of 5 kHz to 10 kHz (high) is where you'll perceive the brightness of a sound. There is little musical content (fundamentals) but it does include most of the harmonics, the sibilance of a voice (the *ssssss* sound), and cymbal sounds. These frequencies add clarity and presence to the overall sound. It would be better to boost in the 6.8 kHz range than in the 10 kHz range. An interesting note is that tape hiss is in the 8 kHz range (you won't run into much tape hiss noise these days, but there you go).
- When a sound seems to hurt or sounds hard to our ears, it's likely we have too much in the 2.5 kHz to 5 kHz frequency range, the upper mid-band. This

is the range where our ears are the most sensitive, especially around 3 kHz to 3.8 kHz. You'll find the screechy part of vocals, snare drum crack, and bass drum attack sounds. Reducing the level in this range might reduce overall clarity, so be scarce with your edits.

- The mid-band itself is two ranges. The first being 800 Hz to 2.5 kHz. Too much of this range can also result in a hard sound. The second range is 315 Hz to 800 Hz and it has lots of upper resonance (drone), especially around the 315 Hz to 500 Hz range.
- Too much of the lower mid-band, 160 Hz to 315 Hz, results in a very muddy/woody sound with lots of lower resonance. Too little from this range can be a loss of warmth. This range contains the fundamentals for the snare drum, toms, bass guitar, tenor vocals, etc.
- Most of our audio energy is found in the sub-bass, lower bass, and the bass bands. This overall range, 40 Hz to 160 Hz, unchecked causes most of the level problems in your mix. **You have a lot of meter level, yet the overall sound isn't very loud.** It's time to cut unwanted lows. Often, an onstage, low frequency resonance rumble (which sometimes results in positive feedback in the microphones) occurs. The reason for this is the interaction between bass guitar, guitar, lower keyboard notes, low toms, and the bass drum resonating with the stage or surrounding items. This is the range we feel more than hear, especially around 40 Hz.

- If the sound system seems to be over driving yet not very loud, your problem could be in the sub-bass and lower bass range. Boost with a sharp Q around 80–100 Hz, check or reduce the 100–200 Hz range, and remove everything below 40 Hz if you want to feel a bit more kick in your sound.
- The frequencies below 40 Hz are not useful (except for precision studio mixing to enhance the low end). Unless you have lots and lots of power amps and a sub-bass speaker system, these frequencies will quickly suck up most of your power. Ahhh … so this is how they're getting all that thump—all those extra car batteries.

The secret of EQ adjustments

At the start of this section, we described the EQ adjustment process:

- EQ space is carving the audio on a track so that the frequency range of that track audibly fits with the other instruments and voices in a mix.
- Each individual instrument or vocal needs to occupy its own space within our range of hearing. You want them to rub elbows but have their own distinct voice and be clearly heard.

The *EQ Space* process can be in many ways a refinement of the *Volume Space* process. You are molding and carving, chiseling away unwanted stone from the statue.

This is all true. But recognize this as only the beginning. The EQ adjustment process is a sharp, powerful tool that can focus a track to help the mix stand out to the listener, or it can quickly kill both the mix and the vibe of the track. Be judicious with EQ moves.

Discover the psychoacoustics behind adjusting the frequencies of two tracks with similar sounds. For example, a synthesizer and the main vocal.

> *You'd think the best approach would be to notch out the matching frequency of the synth at 2 kHz so the vocal frequencies could shine through. But this might diminish a valuable part of the synth's interaction with the vocal. Instead, it might be a different frequency that works in getting the vocal to stand out against the synth. Experiment. The solution to our ears is not always logical!*

The basic practice of using *subtractive* EQ (what you do when you diminish frequencies) is a time-proven method. But the surgical cut of the right frequencies may not be obvious.

And so, this is a world that demands closer inspection. The secret of a superior mix starts with mastering EQ adjustments. Explore the nature of EQ adjusting and psychoacoustics in the *5 Key References* and *Tools* sections.

CHAPTER SEVEN

Effect Space

Here is the icing on the cake. It's another tool to help manage the depth dimension of your mix. Effects can be used to enhance a track or give the illusion of distance. Using effects for both purposes can help your production sparkle. Make a statement with the overt use of effects. Or use effects to help enhance the third dimension of your mix.

For example, combine reverb and volume to give a sense of stage depth. Move a voice or instrument forward and backward on a sound stage by adjusting the levels to your taste.

Each new effect can be its own addition as a track to your mix. You can visualize where the effects might fit into your mix when you've set the basic idea of the *Stereo Pan Space* for the other tracks.

Whenever possible, each voice/instrument should have its own effect. That is not to say everything needs an effect. That is to say, each voice/instrument should have its own *type*, or its own level, of effect when the use of an effect is desired. Everyone's taste will vary.

Even if you choose the same effect for a couple of tracks, each should be separate effect levels and settings. It can be very effective to leave a track dry or with no effect against other tracks that have effects.

The exception is back vocals (again, no hard and fast rules). You will often want to use the same effect and level of effect to help "glue" all the back vocals together.

Remember, effects might alter the perception of, or add to, the original EQ setting of the instrument or voice. Listen to what the effect brings to the overall EQ range of the instrument or voice. You will want to A/B test with/without the effect.

Wet Tracks

Wet tracks are tracks that already have effects printed, or mixed in with, the instrument or voice. Whether these wet tracks are given to you from the original recordist or whether these are wet tracks you've created with effects, you need to use them with *caution*. They can quickly smother what you've built unless you've given prior thought to how you will use them.

Think of them in terms of how they will support what you have built. Decide whether you'll use these tracks as an overt effect or to help build depth perception.

One way of creating an effect is to send a mono or a stereo signal from your source—whether voice or instrument—to the input of the effect plug-in or module (this is a split signal from your dry track).

Route the output to the stereo input return (L or R or both) of an existing track. Or assign the output to an open pair of faders on the console. No matter how you choose to accomplish this, it's important to visualize and remember the signal path. Sometimes it will be best to document it.

You might find it wise to document all of your assignments, for that matter, for your reference. This is especially true if you're not the only mixer using the board (and, I might add, a great way to store and recall different settings for each player).

No matter the signal path you choose, that path is added to the mix and needs adjustment (*Stereo Pan Space, EQ Space,* and *Volume Space* adjustments).

For example, you might find a keyboard recorded with a stereo chorus. You may wish for either channel of the effect to be weaker or sound different from the other.

The bottom line is this: These effect tools are there to help support your tracks. You can make them a part of what makes the song work, or they can quickly detract from your mix. The proactive management of them is a vital fifth key of the *5 Keys*. It's your song, so it's your call.

Effects Won't Save Your Mix

Let me start this section by saying there's no effect trick that's going to save your project. We tend to think otherwise at times, but we would be wrong. There is no better way to present the best face of your project than the best trick of all—*record the source with intention.* Make a note right now, and plan to do further research from the experts. There is no better treatment for your song.

In live music, a church situation and set-ups for recording, you will have the same concerns that need to be addressed with intentionality. Those concerns are proper microphone set-ups and line-level checks before each session. Setting up the microphones and capturing the source with intentionality are critical for:

- receiving the strongest signal from the source,
- receiving the intended timbre of the source, and
- capturing the full range of the source.

When we think something doesn't sound quite right, or is weak, we may think that masking it with reverb or another effect will make that weakness go away. That's not the best solution.

The best solution is to learn the best microphone capturing techniques. Rather than thinking you'd like a certain track or voice to sound a certain way when mixing, consider planning—that is, if you have control over the situation—and capture the voice or instrument with the style you hear for the mix. The better the capture, the more control you have over the vocal or instrument. For example, if the intention of the song is an intimate vocal or finger style acoustic guitar, then perhaps you can choose to record them with a condenser microphone at close range. If the acoustic is to be a loud strummer track, you might choose a dynamic microphone at a greater distance.

There is much to learn from experimenting with microphone capturing techniques. The point regarding *Effects Space* is if you don't capture the voice or instrument with intentionality for the end mix of the song in mind, you'll try to "fix it in the mix" to make it the way you want it to sound, using EQ and effects. And that doesn't always end well.

Make a point to learn more about this subject. Even for the live mix situation, weak signals and bad instrument and microphone cords can wreak havoc on your mind and mixing decisions at the soundboard. Read more about microphone techniques from the experts found in the *5 Key References* section.

The bottom line is this: The fewer effects you use, the clearer your mix. Make your use of them deliberate and intentional. Sometimes adding effects can make a mix or a live instrument sound unnatural. The most natural effects are a good balance of EQ with your chosen effects. Remember that the capture of the room sound with the source is *also* an effect. Here are a few tricks for experimenting with them:

- Send the output of the effect to a separate track.
    - Shape the sound of the track with an EQ plug-in. With the channel fader, you have full control over the volume of the effect.
- EQ the right-side return channel of the effect separately from the left.
    - Or vice versa. When we see a band or an artist live, the room we hear them in makes all the difference in the listening experience. One thing is for sure, we don't always hear the same timbre of the sound in each ear. Play with EQ adjustments in each side of the stereo signal.
- Reverb trail to only one side.
    - Maybe the lead vocal has reverb, but you've sent the tail of the reverb to the right side of the main mix. Or how about the chorus

or flange on only one side? The possibilities are endless. But always consider the full impact of each decision and how it affects the five areas.

- Record an instrument with one or two additional room mics.
    - The holy grail of recording isn't equipment. It's the sound of the room and understanding microphones and source capturing. Take advantage of a great recording space by setting up an additional microphone or two to capture natural reflections. Experiment by capturing sources from several distances. These tracks can provide a very nice natural room reverb.

Enhance Clarity

Can I say it again? Any extra track you create with EQ or effect tricks is like any other track in your mix. It's not enough to just adjust the incoming volume of the effect. These tracks will introduce many frequencies that will interfere with your other instruments and voices.

> *In the end it's all about the sound you like and want to make, it's not necessarily about the science of it.*
> —Jeff Lynne of the Electric Light Orchestra

And a quote from me: "It's your song, so it's your call."

Don't add any effects until you go through the first four *Keys*, unless you have a plan to introduce an effect as a central focus of your project. It's only when you find and remove the main annoying frequencies of each voice and instrument (subtractive EQ) that you should introduce effects. Otherwise, you won't be able to distinguish between the source and the effect on the source. Subtractive EQ might remove more of the source than you intend. Back off the effect and go back and forth between tracks to tailor the EQ to get the instrument or voice to fit right where you want it along the horizon of the *Stereo Pan Space* and the depth of the *Volume Space*.

When you think along these lines, you'll create more clarity in your mix and allow your primary tracks to shine.

CHAPTER EIGHT

The 5 Keys Summary

You're building a 3-D masterpiece. Think of yourself as a sculptor working with a slab of marble. Just as you're trying to release the statue from the slab of marble, so you're trying to release the true essence of the song from the raw material that makes up the song.

Begin to think of your creation, this song, not as a collection of a jillion tracks all somehow affected separately. Think of it as one song. A single product taking on a life of its own. Visualize it as one thing and concentrate on how you can enhance your creation.

Continually consider, rotate through, and adjust each of these *5 Keys* in your mix. Perform surgical EQ cuts with each area in mind. A single EQ cut will affect your entire project.

If you're mixing live, whether at an outdoor concert or in church, your mix moves affect the listening experience—the listener involvement and response—just as any good recording would do. Your goal is removing distractions and getting the listener to hear *only* what you want them to hear.

Begin with the end in mind

Author Steven Covey's advice to "begin with the end in mind" when he talks about *The 7 Habits of Highly Effective People*, applies here, too. What Covey means is if you'll visualize the end product when you begin the construction phase, you'll build each element of your house so they fit together as tight as a jigsaw puzzle.

I've heard it said another way—"design intent". This is the advice given to those who use 3-D CAD software: *Think ahead of time how you want to build the project by incorporating the right elements into your baseline drawing.* If you have the correct design intent, then you won't over- or under-specify each part that you're attempting to build.

You'll always refer back to your base design.

If you know what you want this song to sound like in the final mix, then you'll know how to plan the recording sessions so they'll fit your design intent. For example, you'll be in a position to decide whether you need a keyboard playing throughout the entire song, or to incorporate a cello part, or whatever.

The challenge for the live mixer, since it's not your decision on what instrument or voices are in front of you, is to highlight the main tracks and make the others support the main tracks and the groove and mood of the song.

Stay true to your original decisions

No matter the adjustments and tweaks you make, stay true to your original thoughts about the song. Remember why you made the pick of tracks that made the song work in the

Discovery process. As you listen to the song now, does it still support those original thoughts?

If you change your mind about something, you change your mind. It's your song or your mix. But don't let the "shiny object syndrome" take you away from your original intent.

CHAPTER NINE

Bonus 1 and 2

Take your project to the next level by applying the philosophies of these next two bonuses. Your reward will be a cleaner, more memorable mix. And you'll receive feedback such as this:

> *I sent you an email saying that I missed coming over and recording … We have sold a bunch of demos … They are going like hotcakes. Thanks again for all your help. I have one here with your name on it. Actually, it has our name on it, and we signed it so you will have a memento.*
>
> *Thanks, Scott and Kathryn of PREVAiL*

It's always great to hear that sort of feedback. But their demos were selling like hotcakes because they're a great band. I just helped make sure the recording and mixing didn't get in the way of them showcasing their talents.

The music is the thing. It's all about showcasing the talent, and it's your job not to get in the way of it or cloud it with your mix.

Bonus 1—Find the groove and build the house

Of all the reference material I have ever read regarding mixing and tips on getting a standout mix, there's been no better advice than this: Find the groove and build the house. This section on finding the groove has less to do with the technical aspects of mixing we've covered and more to do with enhancing the mix you've worked so hard to get.

Finding the groove and building the house means you need to understand what the song is about rhythmically. Then reinforce it. You've gone through the *Discovery* exercise of understanding which tracks make the song work. Now, reinforce the song by deciding which tracks are defining the groove and make them shine.

And remember: You have the power to amplify a talented songwriter's task of prosody—the marriage of the music with the lyrics. I first mentioned in Chapter 2 that in songwriting, prosody begins with the lyric. The cadence of the song is dictated by the flow of the words. The message will be conversational and evoke the right emotions. Take the time to recognize the song on this level.

The groove

The songwriters have done a great job of providing lyrical hooks and emotional musical highs. Now it's your job as the mixer to reinforce that prosody.

The rhythm of a well-written song flows along with the emphases of the words as they're spoken in conversation. Emphases of the syllables of the words introduce the intensity of emotion and is a template for the rhythm.

As a mixer, recognize the effort and amplify what the songwriter/artist presents.

One of the best groove support tricks

A well-known engineer received a question about how he goes about getting a mix. One thing he says separates the amateur mixers from the pros is the use of compression.

There's a wide variety of ways to use a compressor, but one useful way is to use the effect to reinforce the kick drum, bass guitar, or other element driving the song rhythmically. This technique has been used very successfully on many records you've heard:

- multing (short for "multiplying", sending the signal to a second track at the same time),
- compressing the mult, and then
- sneaking that track into the mix behind the main track.

Maybe you'd EQ this mult differently. Maybe you'd add a flange effect to it as well. It might be just fine as it is.

I recommend that you learn about compression *only* from the experts in the *5 Key References* section.

Build the House

It's time to build the house. That's just what it sounds like. No structure that is built will last without a firm foundation. As in house building, and life and philosophy in general, so it is in sound recording.

And you've just built a firm foundation. Start where you want the tracks to sit in the mix. Build the mix keeping the five mix rooms of the house in mind. Start with the drums and the bass, or the low-frequency drive instruments of your mix (This too, depends on the song. Maybe a clean electric and a flute drive this song. Just start with what drives this particular song).

Once you have the drivers of the song in place, add that special drive component. Sometimes this drive component might be the lead vocal.

When I mix, I may have that special element in mind, but I'll bring in the rhythm guitar next.

Build your mix upward from here. Bring in the secondary instruments, feature instrument, and back vocals in the order you hear the mix growing and maturing. Build a cradle for the main vocal and feature elements.

> ***Tip:*** *Most of the time, I won't mix in vocals until I build the foundation of the instruments. But bring them into the mix temporarily and get an idea of where you want them to sit in the mix. Then you might take vocals out of your early mix so you can continue to build the foundation.*

Begin with the end in mind. Do whatever makes sense for your production.

Again, it's all about building a lasting and impressionable foundation.

Musicianship and rhythm

Here are a few notes about how rhythm affects studio time and the mixdown process.

The mixer of your songs will want a tight backbeat. If the drums and bass are not in sync (or maybe rhythm guitar and drums), forget about trying to make the rest of the song tight. It begins here. Spend the majority of your studio time making rock-bottom sure of this in tracking. Save money and hours of syncing the tracks later in postproduction.

> *For mixing in church and live bands: Give input to the worship and band leader how you perceive the tightness between bass and drums. You won't have control of this tightness in the live situation, but you can enhance the tightness there by carving the EQ of the kick drum and the bass guitar, panning them center, and adding enhancement from multing as mentioned previously.*

Sure, we have tools today to take care of this job in postproduction, but those fixes can sound artificial and mechanical if not done with care. Try to preserve the human feel of the backbeat. Sure, that's just an opinion, and it depends on the musical style.

Regardless of the tools available in the studio, your spending will increase beyond your budget due to added mixdown hours if you're not prepared.

Just a small aside for those of you who are new to recording (this advice is for playing live or in a studio):

> *Lay back on the beat and don't rush it. Lay back on the beat and emphasize the beat strongly when it's called for. Rock on the 1's and 3's and Roll on the 2's and 4's.*

But … don't skim over the 2's and 4's in an effort to lay into the 1's and 3's. Those beats are just as important.

Respect the space of the 2 and the 4. This is how you get a great groove going. Sure, it depends on the type of rhythm of the song and the different emphases you can place on different beats. No matter the type of music, think like you are a time piece, a metronome.

I give more details on these topics and more in my free e-book, *How to Prepare for the Studio* at https://songs4god.net/how-to-prepare-for-the-studio.

Bonus 2—Dynamic mixing

The next level of mixing, both live and in the studio, is moving faders, knobs, and parameters during the song. Placing this tool in your tool belt takes your mixing skills to the next level. You can automate these moves with a DAW. As the audio playback and recording passes the timeline, there are marker events recorded and played back in the DAW song file. They're recorded as digital events by the audio management software. All functions of a DAW, from command events to plug-in parameters on effects, to channel volume and gain, to LR panning, to playback track gain and EQ adjustments, are programmable.

DAWs can be programmed for live shows, depending on the complexity. Most moves, though, are made manually by the operator. This is, of course, always suited to taste. For church sound reinforcement, dynamic mixing using manual moves is smart planning. Even with churches using background tracks, the live players are the channels to be concerned with for dynamic moves.

For example, if a lead guitar part comes, a boost to their channel volume works well, then back down to the previous level. "Riding" the volume fader of the lead vocalist channel can work wonders for the mix, as you study the vocalist and the song for when volume dynamics are needed (You may need to boost quieter passages. This technique is also important if your vocalist moves the microphone too much or doesn't know how to work a microphone). Riding the volume fader of the lead vocal in the studio when making records is a common practice. That can be automated, if desired.

Another situation in the live environment is to widen the stereo panning during a keyboard passage. There are many possibilities. I will say that in the church live environment, I've found that riding faders on vocals and guitars is more common than any other dynamic movement.

Go Forth and Mix

I can't say it any better than one of the most prolific audio mixers on the planet:

> *"As long as you can defend your mix positions with logical, identifiable, and demonstrable arguments, as long as you think about your positions and can justify them to your clients how your decisions best serve the song, then you're in the right. Be confident in the here and now, and if you ultimately determine that your confidence was unfounded, learn from your mistakes and do better next time with even more confidence."*
>
> **Mixerman,** gold and platinum award winning mixer, producer, and author

Think about this: Who is your client? The answer's obvious if you're a hired hand. **If you're mixing for yourself, put your client hat on. Now mix for this client. If you're mixing for the congregation and for the glory of God, give them and him nothing short of your best effort.**

You're creating a philosophy of mixdown by gathering information, making mixes, making *calculated* and *intuitive* judgment calls, being right, and sometimes being wrong. You're always learning. But you're constantly forming your own mixdown philosophy.

Make these *5 Keys* a part of your mixing future.

5 KEY REFERENCES

You'll find a wealth of information about recording, mixing, and mastering on the web. You'll also find a lot of information that's not conducive to good audio mixing education. That's just the nature of the beast. And there are varied opinions in between.

How do you decipher *your* truth about them? You test everything. Your mind has to be the best judge. Take everything you read with a grain of salt. Test everything you read against actual experience. Confirm what you read against reputable sources and what you know to be true. Build your own library of knowledge.

Be wary of cookie-cutter solutions. What works for me in my situation might not work for you.

Key Reference 1

Bobby Owsinski is one of the best-selling authors in the music industry with over twenty books, including *The Mixing Engineer's Handbook, The Recording Engineer's Handbook,* and *The Mastering Engineer's Handbook.* Bobby is a great communicator that has spent his life sharing his knowledge and creating a business model to sell his knowledge. He's not only a smart businessman, he shares that knowledge earnestly and clearly. He simplifies the most complicated studio moves.

You could spend a lifetime reading his books, but he's also a prolific blogger. Go to https://bobbyowsinskiblog.com where you'll find countless articles on the music business, all things studio and social media, and over five very detailed online courses covering every aspect imaginable in the mixing world. He also created the "Hit Maker's Club", a social networking membership for producers and mixers of every level. Here is Bobby describing his goal for the club:

> *I want you to feel the confidence that comes with knowing that the gap between you and the experts is no longer as big as you thought.*

Key Reference 2

Warren Huart is a musician, producer, composer, and recording engineer most known for his role as a multi-platinum producer for The Fray, Ace Frehley, Aerosmith, Daniel Powter, Marc Broussard, Korn, and Howie Day, among many others.

Warren designed the membership program called "Produce Like a Pro Academy" with a YouTube subscription of over 600,000 people worldwide. This program is an amazing golden-vault collection of over fifty sets of multi-tracks, thirteen full mixing courses from industry professionals, a twelve-week mastering course, direct feedback and support from an online pro-audio community and incredible interviews with the industry's brightest minds and talented mixers and musicians. Learn more about Warren at https://producelikeapro.com.

I recently came across a brilliant video of his where he shares his mix philosophy, the 5 Rules of Mixing. This is epic information for the new as well as seasoned mixer, spoken with elegance and clarity:

https://www.youtube.com/watch?v=NE1Ov-ObLv0

Key Reference 3

Eric Sarafin (better known on the internet as Mixerman) is a Gold, Platinum, and Multi-Platinum record producer, recordist and mixer. He has the mixing credits of many major label projects, including the artists Foreigner, Ben Harper, Pharcyde, Amy Grant, Bare Naked Ladies, and Hilary Duff, just to name a few.

That alone is an impressive enough reason to read and listen to what he writes and records. Mixerman is a passionate writer that has shared his love for recording and mixing like no other experienced professional, and we are much wiser for it.

Mixerman has been quite an influence on me. He not only has a mix philosophy, but also a philosophy of recording, producing, and studio time management.

There are two of his written works I find particularly important for gaining knowledge about mixing:

- *Zen and the Art of Mixing*
- *Musician's Survival Guide to a Killer Record*

In particular, you'll discover his holistic philosophy on the entire process of recording and mixing in the *Survival Guide*. "Valuable" is not a strong enough word.

He's been accused of "giving away the farm" as far as insider industry knowledge. He may have, but there is no written word more beautiful than the truth. He brings that. These books are *much* more than tips and experience. This is the next level of knowledge you should seek to add to your mixing philosophy.

Find an incredible amount of knowledge sharing and products by visiting his website, https://mixerman.net/.

Key Reference 4

Mike Stavrou has been mixing music in the US, UK, and Australia for 40 years. You know he knows what he's talking about when his book, *Mixing With Your Mind: Closely Guarded Secrets of Sound Balance Engineering REVEALED*—has a foreword from Sir George Martin (producer of the Beatles).

Stav worked as an intern for Sir George before branching out on his own. You'll find Mike's tips, tricks, and philosophies to be *at the next level.* That's because Stav created his own recording and mixing philosophy of "Maximum Illusion, Minimum Voltage".

I can't say any more. Just know that Stav is all about psychoacoustics and using the brain to mix a record. I am blown away by the number of moments when the "penny dropped" as he shared his method with examples in this book. Must. Have. Book.

Let Stav reveal his secrets of using your mind and the benefits of understanding psychoacoustics by picking up a copy today. You won't find this book at Amazon, except second-hand copies that are more expensive than the original! It only ships from the author direct to you: http://mixingwithyourmind.com/

Key Reference 5

John Vestman owns and operates a mastering studio in Orange County, CA. John's clients and artists include gold, platinum and Grammy winners spanning a continuous 35+ year career. He has a massive list of accomplishments as an audio engineer, author, pro-audio equipment designer, and industrial/film score production engineer. His website, http://www.johnvestman.com, contains a vault of recording, mixing, and mastering information you'll want to sink your teeth into! There's a wealth of information and microphone tips, recording tips, everything you wanted to know and ask about digital versus analog, etc.

He offers a stem-mastering philosophy, where he processes groups of tracks in the mastering process. John mastered a record for me with that philosophy, and it turned out great. He has a wealth of experience. I guarantee that you will bookmark this site! Also see https://www.linkedin.com/in/john-vestman-8a504610/

TOOLS

Audio Technician as a member of the worship team

https://stephenrobertcass.com/books

Establishing a Culture of Lead Worshipers: How to Build a Worship Team. How the sound volunteer, or the paid position, is a member of the worship team. And not just on Sundays.

https://worshipleader.com/technology/bridging-the-gap-from-worship-leader-to-sound-tech/

https://www.prosoundweb.com/who-defines-good-sound-the-sound-tech-as-an-equal-worship-team-member/

https://www.pursuegodnetwork.org/how-worship-leader-sound-tech-and-media-tech-roles-work-together/

https://www.behindthemixer.com/passing-out-sound-tech-job-description/

https://www.worshipteamcoach.com/tech/011-how-to-lead-and-learn-from-your-sound-tech/

Church Audio Training

https://churchsoundu.thinkific.com/courses/101-complete

https://www.churchsoundbootcamp.com/

https://courses.worshipsoundguy.com/

https://easychurchtech.com/free-audio-training-for-churches/

https://prosoundweb.com

Mixing and Recording Education

https://www.soundcraft.com/ui24-software-demo/mixer.html

https://creativesoundlab.tv

https://mixerman.net

https://bobbyowsinskiblog.com

http://johnvestman.com

https://producelikeapro.com

http://mixingwithyourmind.com

Acoustic Treatment for Home Studios

https://ehomerecordingstudio.com/acoustic-treatment-101/

https://www.izotope.com/en/learn/acoustic-room-treatment-on-a-low-budget.html

https://www.soundonsound.com/techniques/studio-sos-guide-monitoring-acoustic-treatment

Song Construction for the Audio Tech and for Making Demos

https://stephenrobertcass.com/books

Research "prosody" in *Fishing in Church: How to Be a Congregational Songwriter.*

Basic song construction steps to identify a strong song.

https://stephenrobertcass.com/9-steps-to-build-a-memorable-song

EQ Adjustment Education

https://www.audio-issues.com/music-mixing/all-the-eq-information-youll-ever-need/

https://blog.landr.com/eq-basics-everything-musicians-need-know-eq/

https://www.presonus.com/learn/technical-articles/What-Is-a-Parametric-Eq

https://www.musicianonamission.com/approach-equalization-two-types-eq/

Recording Studio Prep

https://songs4god.net/how-to-prepare-for-the-studio/

https://www.musicradar.com/tuition/tech/10-things-you-should-do-before-you-go-into-a-pro-recording-studio-168586

https://www.nolitastudios.com/blog/2019/4/23/6-ways-to-prepare-for-the-recording-studio

https://blog.sonicbids.com/5-tips-every-musician-needs-to-better-prepare-for-a-recording-session

https://audient.com/tutorial/5-steps-prepare-band-studio/

ABOUT THE AUTHOR

Stephen Robert Cass is a hack golfer who aims for high mediocrity. When not on the course, he's known for:

- ❑ 50+ years as a worship musician and team member,
- ❑ 14 years as a worship leader,
- ❑ 26 years as a published Christian songwriter,
- ❑ 15 album projects, whether solo, produced or musician credits.
- ❑ 70+ worship song titles found at Church Copyright Licensing, International under the Solid Walnut Music catalog and Stephen Robert Cass.

Solid Walnut Music has given away music CDs to Christian radio stations all over the world: the U.S., Canada, Mexico, Australia, England, Ireland, Russia, Ethiopia, Bulgaria, Italy, South Africa, South Korea, and Israel.

Contact Steve@songs4god.net.

* * *

Made in the USA
Monee, IL
04 November 2026

67378972R00056